# Panther Tank

Written by David Doyle

In Action®

Squadron/Signal Publications

Art by Vincenzo Auletta

(Front Cover) With its heavy sloping armor, powerful 7.5cm main gun, and wide tracks, the Panther was one of the best WWII medium tanks. About 6,000 were made during the war, but few now remain. One survivor is this operational Panther Ausf. A built by M.N.H. in June 1944 and owned by the Musée des Blindés, Saumur, France. (Massimo Foti)

(Back Cover) Even more heavily armed and armored than the Panther tank was its tank-destroyer derivative, the Jagdpanther. This operational example, originally issued to the 116th Panzer Division, is owned and restored by the Weald Foundation in the UK, after decades of being reduced to a hulk as a range target in England. (John Blackman)

# About the In Action® Series

*In Action®* books, despite the title of the genre, are books that trace the development of a single type of aircraft, armored vehicle, or ship from prototype to the final production variant. Experimental or "one-off" variants can also be included. Our first *In Action®* book was printed in 1971.

**ISBN 978-0-89747-831-1**

**Proudly printed in the U.S.A.**



**1115 Crowley Drive, Carrollton, TX 75006-1312 U.S.A.**



# Military/Combat Photographs and Snapshots

If you have any photos of aircraft, armor, soldiers, or ships of any nation, particularly wartime snapshots, please share them with us and help make Squadron/Signal's books all the more interesting and complete in the future. Any photograph sent to us will be copied and returned as requested. Electronic images are preferred. The donor will be fully credited for any photos used. Please send them to:

**Squadron/Signal Publications**
**1115 Crowley Drive**
**Carrollton, TX 75006-1312 U.S.A.**
**www.Squadron.com**

(Title Page) Panther Ausf. G tanks pass each other on a road in the Alsace-Lorraine area of what is now France in September 1944. Notably, the vehicle advancing toward the camera has tracks that lack chevrons (*Gleitschutzpickeln*). (Bundesarchiv)

# Acknowledgments

Beyond the invaluable help provided by the staffs of the National Archives, the Bundesarchiv, and the Patton Museum, I am indebted to Tom Kailbourn, the late Walter Spielberger, Thomas Anderson, Dana Bell, Scott Taylor, Hans-Heiri Stapfer, John Blackman, and Massimo Foti. Their generous and skillful assistance adds immensely to the quality of this volume. On this project and all aspects of my life, I am especially blessed to have the faithful and tireless help of my wonderful wife Denise, who has scanned countless photos and documents for this and numerous other books.

# Introduction

Though not as celebrated in literature (or propaganda) as the much feared Tiger, the Panther tank was arguably the more effective of Germany's two World War II-era "cats." Sometimes known as the Panzerkampfwagen V, the Panther was introduced in 1943 as Germany's new medium tank, intended to replace both the Panzer III and Panzer IV, both of which proved outclassed by the Soviet T-34 and KV-1. The tank had excellent firepower and protection for its day, but early Panthers especially were plagued with mechanical issues, many of which were the result of the vehicles being rushed into production and into the field where they were manned by troops who lacked adequate training on the tanks – all in a feverish effort to stem the Soviet counteroffensive in 1943.

While some issues, such as troop training, were corrected – at least in the short term, some of the mechanical issues, notably problems with the transmission, continued to plague the Panther to the end of the war. These issues were compounded by the incessant Allied bombing not only of the tank assembly plants, but also of the factories that made key parts and subassemblies. A further problem for the Panther arose as a result of the use of slave labor to manufacture the tanks and their components. Intended in part to lower the 117,100 Reichmark vehicle cost, the use of hostile prisoners in industry resulted in frequent instances of sabotage that, a 1947 study showed, were effective.

With its sloping armor, both to deflect shot and to increase effective thickness, even 75 years later the Panther still has a modern appearance. Nevertheless, the tank failed to live up to its potential. Originally designed as a 30-ton vehicle, with an engine and transmission sized accordingly, as typical of German tank designs, the Panther's weight soon began to climb, ultimately soaring to over 45 tons, which astoundingly is nearly the original design weight of the massive Tiger tank.

**As pictured in this cyanotype, the Soviet T-34 was directly responsible for the creation of the Panther. This then-new Soviet tank outclassed the German Panzer III and Panzer IV vehicles, which previously had the decided advantage on the battlefield. The T-34 had better firepower, protection and mobility, and the German forces demanded a tank that not only equaled the T-34, but bettered it. (Bundesarchiv)**

**This is the second of two Panther prototypes produced by M.A.N. in the fall of 1942. The first prototype was an automotive test vehicle and lacked a turret. Known as *Versuchs-Panthers,* the drive sprocket of the prototypes differed considerably from those used on production vehicles. (Thomas Anderson collection)**

**M.A.N. and Daimler-Benz submitted proposals for a new tank based on lessons learned with the Soviet T-34. The Daimler-Benz VK 3002 proposal, a model of which is shown here, obviously owed more to the T-34's design than M.A.N.'s winning proposal.**

## Specifications

| | Ausf. D | Ausf. A | Ausf. G |
|---|---|---|---|
| Length | 8.86 meters | 8.86 meters | 8.86 meters |
| Width | 3.27 meters | 3.42 meters | 3.42 meters |
| Height | 2.99 meters | 3.10 meters | 3.10 meters |
| Weight | 44.8 metric tons | 45.5 metric tons | 45.5 metric tons |
| Fuel Capacity | 730 Liters | 730 Liters | 700 Liters |
| Maximum Speed | 55km/hr | 55km/hr | 46km/hr |
| Range, On Road | 200km | 200km | 200km |
| Range, Cross Country | 100km | 100km | 100km |
| Crew | 5 | 5 | 5 |
| Communications | Fu 5 and Fu 2, Intercom | Fu 5 and Fu 2, Intercom | Fu 5 and Fu 2, Intercom |
| Chassis Numbers<br>M.A.N.<br>Daimler-Benz<br>Henschel<br>M.N.H.<br>Demag | <br>210001-210254<br>211001-211250<br>212001-212130<br>213001-213220<br>--- | <br>210255-210899<br>151901-152575<br>---<br>154801-155630<br>158101-158150 | <br>120301-121443 (apx.)<br>124301-125304 (apx.)<br>---<br>128301-129114<br>--- |
| | | | |
| Weapon, Main | 7.5cm KwK 42 L/70 | 7.5cm KwK 42 L/70 | 7.5cm KwK 42 L/70 |
| Elevation | -8° to +18° | -8° to +18° | -8° to +18° |
| Weapon, Secondary | 7.92mm MG 34 | 7.92mm MG 34 | 7.92mm MG 34 |
| In Hull | 7.92mm MG 34 | 7.92mm MG 34 | 7.92mm MG 34 |
| Ammo Stowage, Main | 79 rounds | 79 rounds | 82 rounds |
| Ammo Stowage, Secondary | 5,100 rounds | 5,100 rounds | 4,800 rounds |
| | | | |
| Armor, Glacis | 80mm / 55° | 80mm / 55° | 80mm / 55° |
| Armor, Hull Side | 40mm / 0° | 40mm / 0° | 40mm / 0° |
| Gun Mantlet | 100mm | 100mm | 100-110mm |
| Turret Front | 100mm / 12° | 100mm / 12° | 100mm / 12° |
| Turret Side/Rear | 45mm / 25° | 45mm / 25° | 45mm / 25° |
| | | | |
| Engine Make, Config. | Maybach V-12 | Maybach V-12 | Maybach V-12 |
| Engine Model | HL230 TRM P30 | HL230 TRM P30 | HL230 TRM P30 |
| Engine Displacement | 23 Liters | 23 Liters | 23 Liters |
| Engine Horsepower | 650 @ 3,000 R.P.M. | 700 @ 3,000 R.P.M. | 600 @ 2,500 R.P.M. |

### Panther Ausf. D

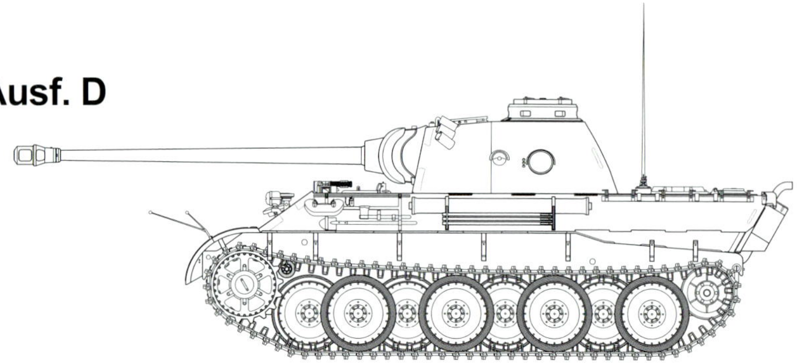

### Panzerbefehlswagen Panther Ausf. D

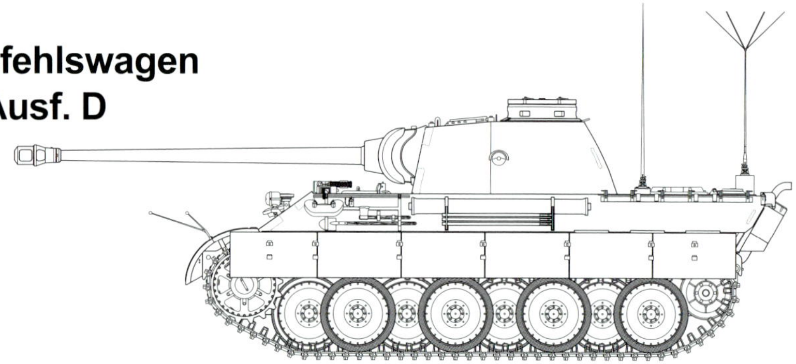

### Panther Ausf. A

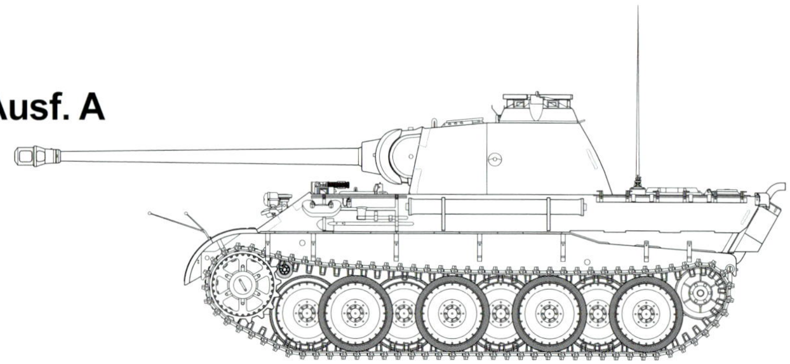

### Panther Ausf. G

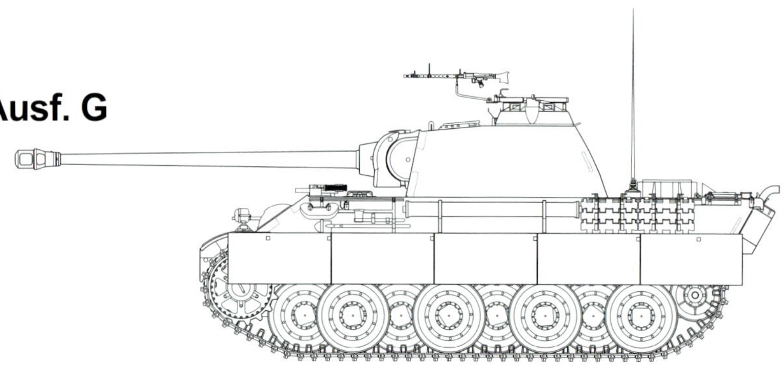

# Ausf. D

The first production model of the Panther tank was the Ausführung D (Ausf. D) ("model D"), the first of which were completed in January 1943, following a brief, nine-month development and initial production period. The Panther Ausf. D made use of an existing but modified Rheinmetall turret, mounting a powerful 7.5cm Kampfwagenkanone KwK 42 L/70 gun. The hull incorporated sloped armor to increase the protective properties of the plate. The driver had a vision port with a hinged cover: a feature that carried over to the next production model, the Ausf. A. Seen here are Panther Ausf. Ds in overall *Dunkelgelb* (dark yellow) camouflage paint, being prepared for shipment by railroad flatcars in 1943. (Bundesarchiv)

**Workers on the Henschel & Sohn assembly line in Kassel appear oblivious to the several tons of Panther Ausf. D hull passing over their heads in June 1943. Once welded together, the hull was machined for installation of the various components prior to beginning its trip down the final assembly line (Bundesarchiv)**



The first model of the Panther to begin coming off the assembly line was the Ausführung, or model, D. One thousand of the new tanks had been ordered from Maschinenfabrik Augsburg-Nürnberg (M.A.N.), Daimler-Benz, Maschinenfabrik Niedersachsen-Hannover (M.N.H.) and Henschel und Sohn in Kassel. Interestingly, Henschel would build only chassis, utilizing turrets produced by Wegman, also in Kassel. The other firms would produce both chassis and turrets.

Even before the first tank on the initial 1,000-tank order was completed, that order had been reduced to 850 units. In part due to the diversion of some chassis for Bergepanther production, ultimately only 842 Panther Ausf. D were completed as combat tanks.

Despite Adolf Hitler's insistence on quick production, the first Panthers did not begin to roll off the assembly lines until January 1943, when the four firms involved completed 26 vehicles. Once production was underway, the combat debut was further delayed by the necessity of rebuilding the tanks due to various problems discovered even as production was underway – problems notably with the engines.

Ultimately, Panzer-Abteilung 51 on the Eastern Front was issued 96 of the new tanks on 24-25 June 1943. A further 96 plus two Bergepanthers were issued to Panzer-Abteilung 52, also on the Eastern Front, a few days later.

Then just a few days after that, on 5 July 1943, 184 of these vehicles were committed to action in the Battle of Kursk. As revealed in a report by Generaloberst Heinz Guderian – then Inspekteur der Panzertruppen – to the Chief of Staff, within two days, only 40 Panthers remained operational; after five days this figure had dropped to 10 vehicles. Twenty-five were total losses, while 56 had been damaged in combat and 44 were known to have broken down, the remainder littered the battlefield, their condition unknown.

**Factory and military representatives pose next to Panther Ausf. D chassis number 210111 at the Maschinenfabrik-Augsburg-Nürnberg (M.A.N.) factory after it was remanufactured with the latest improvements in June 1943. "KV" stood for combat ready *(Kriegsverwendungsfähig)*, and this tank was the first tank to come out of the program. (Hans-Heiri Stapfer collection)**

A Panther Ausf. D undergoes analysis at a proving ground. The light-colored rectangle on the upper right corner of the glacis contains the number 1A-0805. This vehicle was completed after June 1943, as indicated by the presence of only one headlight. (Patton Museum)

In August or September 1943, the crew of a Panther Ausf. D makes repairs to the final drive and the left track of their vehicle at a site near the Mius River in the Donets Basin in the southern USSR. Track repairs and changes typically were very arduous tasks. (Bundesarchiv)

A member of the crew of a Panther Ausf. D poses in the rear hatch of the turret, somewhere in northern France in the fall of 1943. This photo provides a rare, close-up study of Zimmerit as applied to a Panther Ausf. D. Tactical number 425 is painted on the cupola. (Bundesarchiv)

Crews load Panther Ausf. Ds onto flatcars for shipment to the front. Protective bundles of sticks have been wrapped around the 7.5cm gun barrels on the first two tanks. (NARA)

A Panther Ausf. D of Panzergrenadier Division "Großdeutschland" provides fire support on the Russian southern front in 1944. An armored hood for the telescoping snorkel, an Ausf. D feature, is on the engine deck to the front of the right exhaust pipe. (Bundesarchiv)

The crew of a Panther Ausf. D pauses on an Eastern Front road march in June or July 1943. Between the two storage boxes on the engine deck is an armored hood that covers the opening for a telescoping snorkel for deep-water fording. (Patton Museum)

Two crewmen change a wheel on the left side of a Panther Ausf. D. On the bottom of the sponson are hangers for the armored skirts, which feature upturned tabs on the ends that were fastened through openings in the skirt panels. (Bundesarchiv)

In a view from about the same time and place as the preceding image, the commander of a Panther Ausf. D scans the sky with his binoculars in search of Soviet aircraft. The cylinder on the sponson contains a bore-cleaning brush and staff. (Bundesarchiv)

A Panther Ausf. D emerges from a forest while negotiating a muddy trail in the USSR during December 1943. The site was outside of Kirovograd (now Kropyvnytskyi), in central Ukraine. The letterbox cover for the bow machine gun shows as an area that is relatively smoother than the mud-spattered glacis surrounding it. Faintly visible above the dual sight apertures on the left side of the turret is a curved rain gutter, introduced to Ausf. D production in June 1943. (Bundesarchiv)

A Panther advances along a dusty road in the Soviet Union during the warm months of 1943. The drum-type commander's cupola identifies this as an Ausf. D tank. The dark, oblong feature on the right rear of the turret is the interlock with the rear plate of the turret. (Bundesarchiv)

A pair of Panthers Ausf. D is parked along a road in a forest in the northwestern Soviet Union in the winter of 1943-1944, fir branches providing camouflage. The tank on the left, marked number 438 on the turret, may be identifed as an Ausf. D instead of an A by the shape of the semicircular extension of the turret front behind the mantlet. (Bundesarchiv)

A German machine gunner with an MG 34 mans a foxhole next to a Panther Ausf. D. In the distance is a Sturmgeschütz III (StuG III). The Panther has a faded coat of whitewash. A dust cover is fitted over the 7.5cm gun muzzle. There is Zimmerit on the skirts.

The crews of two Panther Ausf. Ds take a break on a road march. The turret of the rear tank is traversed to the rear. On the right front corner of the turret roof of the closer Panther are spare track links, atop which sits a storage box with its lid ajar. (Bundesarchiv)

An early-production Panther Ausf. D with smoke-grenade dischargers on the turret rests along the side of a road in the southern part of the Eastern Front in September 1943. A dark camouflage scheme has been applied at various places over the *Dunkelgelb* base color. A heat shield is visible on the inboard side of the left stowage bin on the rear of the hull. What appears to be a ladder is stored on the right side of the vehicle, and an improvised step made of a bent steel rod is attached to the next-to-rear skirt panel. (Bundesarchiv)

# Panther Ausf. D

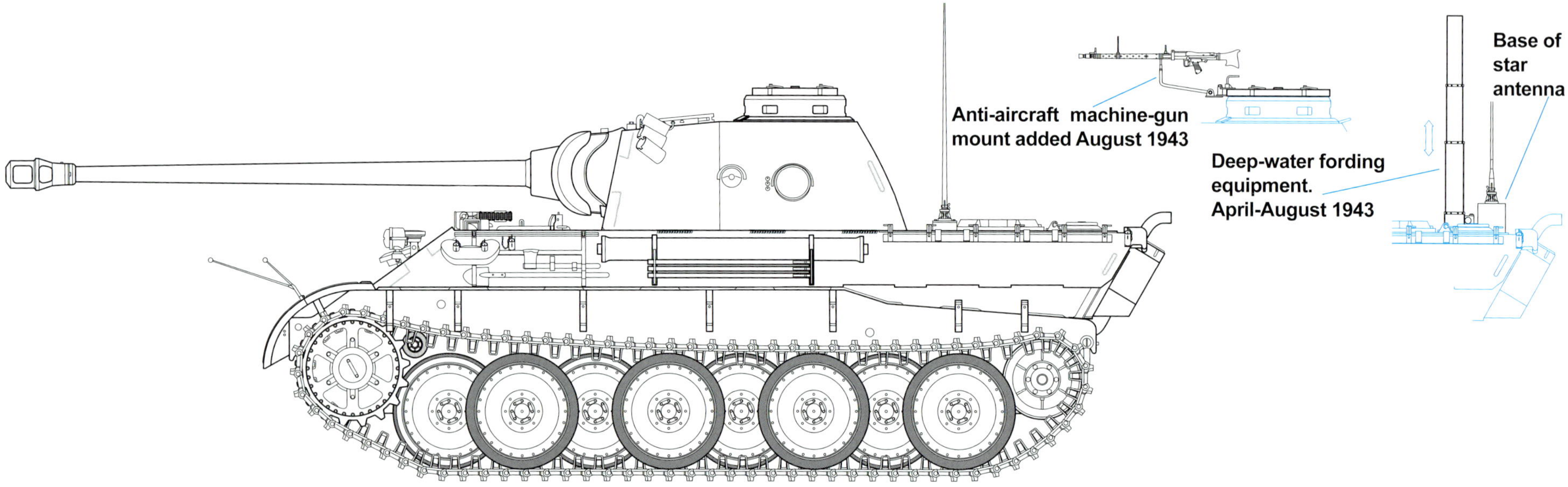


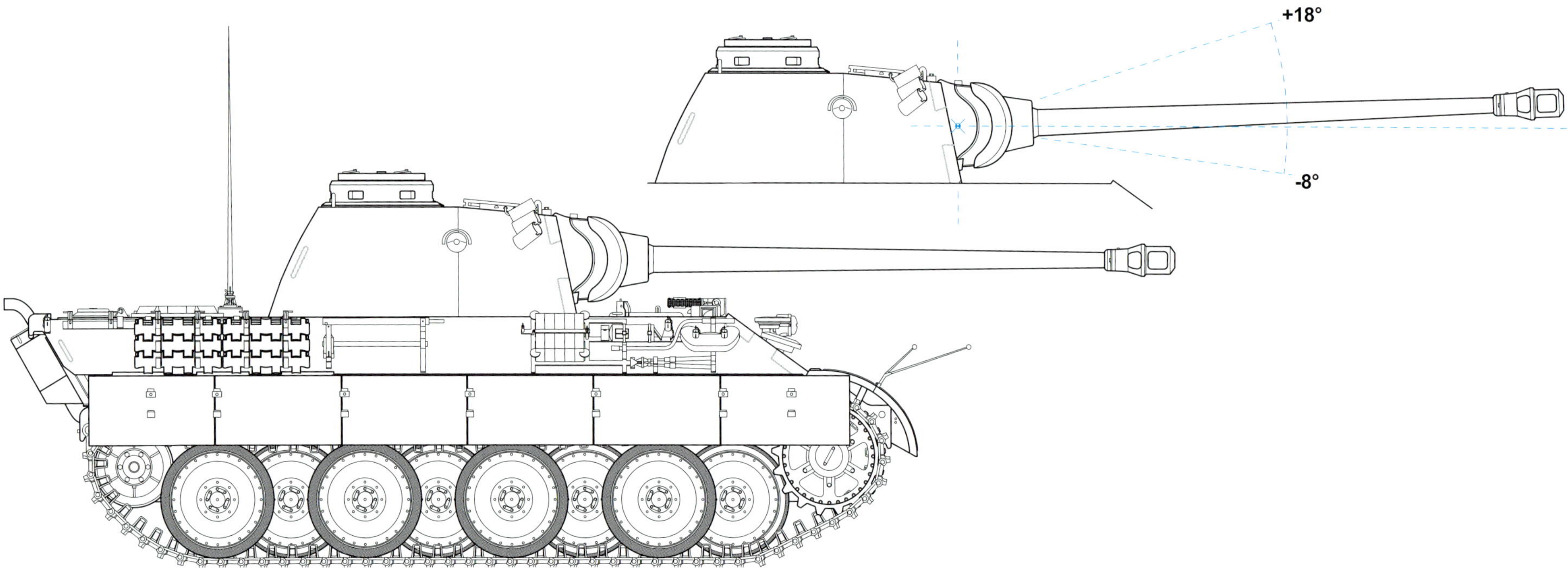

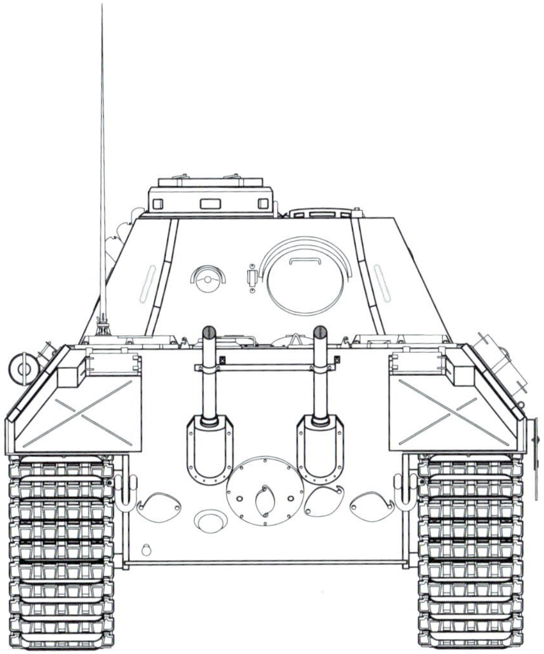

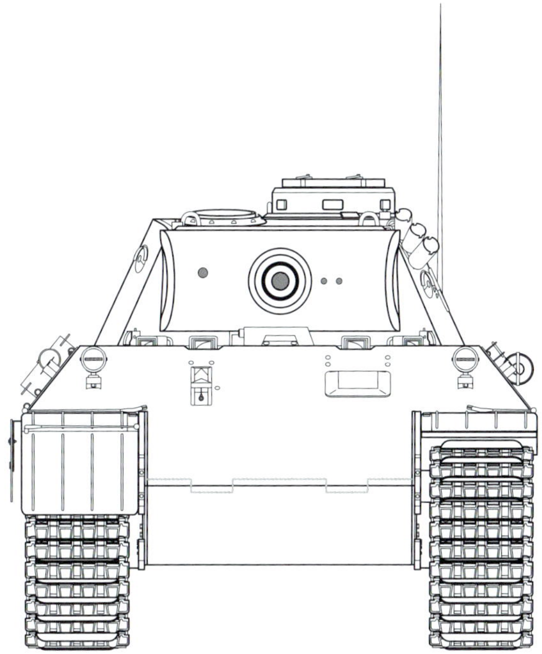

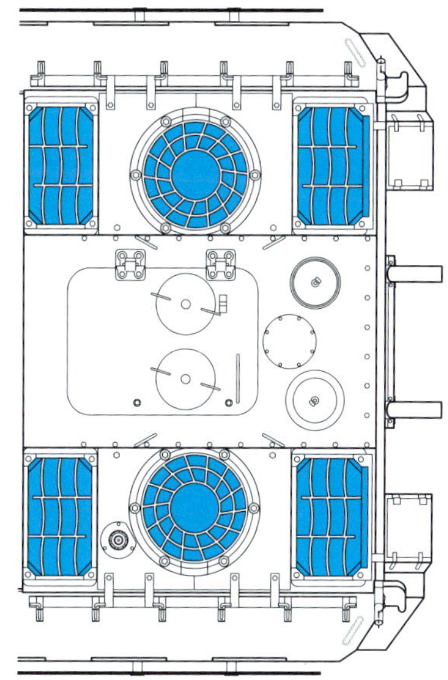

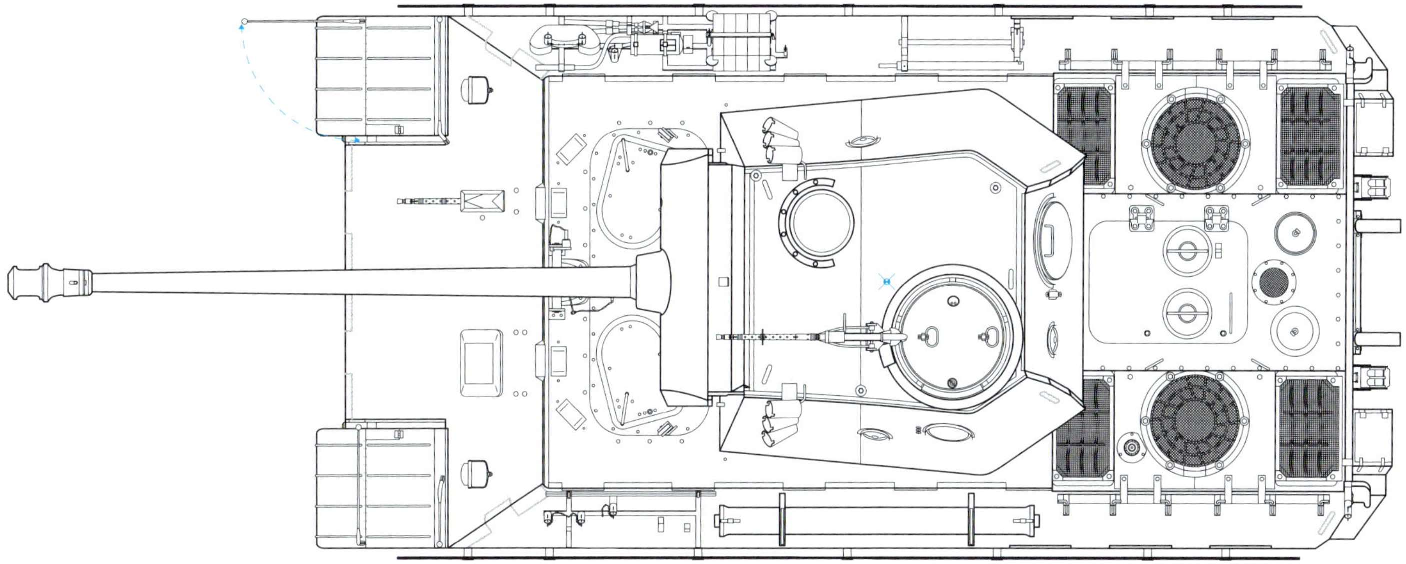

An early Panther Ausf. D from the 4th Platoon, 4th Company, Panzer-Regiment Großdeutschland bears colors and markings used during fighting near Karachev, Russia, in 1943.

There are no smoke dischargers on the turret of this late-production Panther Ausf. D from an unidentified unit.

# Ausf. A

With production spanning from August 1943 to July 1944, the Panther Ausf. A essentially consisted of the Panther Ausf. D chassis coupled with an improved turret. This example built by Maschinenfabrik Augsburg Nürnberg (M.A.N.), on whose factory grounds it was photographed, has the cast cupola and Zimmerit, the latter being applied to the Ausf. A in late August or early September 1943. Although the drum-shaped cupola of the Ausf. D seems to have been installed on the first few Ausf. A tanks, by September 1943 it had been replaced by a new, cast cupola with seven periscopes: the cupola with which this tank is equipped. For the first few months of production by M.A.N., the Ausf. A shared the Ausf. D's "letterbox" cover for the bow MG 34 machine-gun aperture and binocular gun-sight apertures in the mantlet. The mantlet, however, was now noticeably wider. Three pistol ports initially were fitted on the turret, but these were omitted beginning in December 1943 in favor of the roof-mounted *Nahverteidigungswaffe* or "closs-defense weapon." The *Nahverteidigungswaffe* was originally developed as as smoke grenade launcher, but a time-delay explosive round was developed for it, allowing it to defend an area of 20-30 feet around the tank. (Walter Spielberger)

Except for the use of a cast cupola instead of a drum-shaped cupola, the M.A.N. Panther Ausf. A featured here and the two photos below appears from the rear virtually identical to a late-production Ausf. D. The jack is stored below the exhausts. (Walter Spielberger)

The second model of the Panther to enter production was the Panther Ausf. A, which began rolling off of the assembly lines in August 1943. While many Panthers had been damaged or lost at Kursk, much of the fault was placed not on the tank design as much as on inadequate training, and even more significantly, on a lack of proper field trials before committing the vehicles to such a major battle. It is worthwhile to mention that while the Panther suffered considerably, the Germans claimed 140 Soviet tanks in return – including knocking out a T-34 at a whopping 3,000 meters.

Not surprisingly then, the Panther A differed little from the Panther Ausf. D, at least externally. Even internally, initially the hulls were identical. The HL230TRM V-12 engine, which had replaced the HL210 after the first 250 Panther Ausf. D were produced, continued to be used on the Ausf. A. Indeed, the HL230 would continue to be the engine found in all subsequent Panthers, although plans did call for the introduction of the fuel-injected HL234, but the war ended before that engine model entered production.

The Ausf. D turret, however, was extensively redesigned for the Ausf. A. Though the shape of the turret was little changed, most of the mechanics of the turret were revised. Initially, the turret even retained the same type of drum cupola as found on the Ausf. D, however a cast commander's cupola with seven periscopes quickly replaced the drum-style unit.

In late November or early December the hull was changed, with the ball-mount machine gun replacing the so-called letterbox machine gun firing port on the glacis.

A particular feature that differentiated the Pather Ausf. A from the Ausf. D was the shape of the D-shaped armor extensions on the front of the turret that snugly fit to the rear of the mantlet. (Walter Spielberger)

As seen on the same Panther Ausf. A in the photo on page 17, the monocular gun sight aperture on the mantlet to the left side of the 7.5cm gun and the ball-type bow machine gun mount, modifications dating to November 1943, were visible departures from the Ausf. D. (Walter Spielberger)

At the front of the side of the hull of this Ausf. A chassis under construction at the M.A.N. factory is the mounting for the left final-drive assembly. On the side armor above the mounting is the elongated tab where it interlocks with the glacis. Just below the mounting and at intervals on the lower part of the side are access plates for the torsion bar mountings. Above the second access plate are five threaded studs for holding a bump-stop. The first, second, and seventh road-wheel arms are reinforced. (Patton Museum).

On another Panther Ausf. A chassis on the M.A.N. assembly line, the left final drive assembly and most of the road wheels have been installed. The small roller with the black rubber tire attached to the final drive was intended to prevent the track from bunching up on the sprocket when the vehicle was being driven in reverse. On the side of the lower hull, Zimmerit with a vertically ridged texture has been applied over the light-colored material seen in the preceding photograph, that material apparently being an initial, fairly smooth coat of Zimmerit to give the final coat some tooth to adhere to. The textured Zimmerit was kept away from the suspension and final drive. (Patton Museum)

The sprocket, idler, and all of the road wheels are installed on this Panther Ausf. A chassis at the M.A.N. factory. Also present are the tool rack toward the front of the sponson and the tubular container for the bore-cleaning brush and staff on the center of the sponson. Vertically ribbed Zimmerit is on the sides of the lower hull and sponson. (Patton Museum)

Zimmerit has been applied to the glacis of this Panther Ausf. A at the M.A.N. factory. The pattern of the Zimmerit features grids several inches square, within which are vertical ridges. Zimmerit is absent in the ball mount and driver's visor. (Patton Museum)

In the front interior of a Panther Ausf. A chassis under construction, the opening for the driver's visor and the ball mount are toward the top, steering brakes and the right front shock absorber are to the sides, and torsion bars are at the bottom. (Patton Museum)

The five fuel tanks are in place on this Panther Ausf. A. Each sponson holds a tank whose outboard side is angled to match the armor's slope, with another tank below it. In the rear of the center bay is a tank with a bowl-shaped fuel filler atop it. (Patton Museum)

The fan and water-cooler grilles on the engine deck of Panzerbefehlswagen Panther Ausf. A, chassis number 210651, have screens to repel hand grenades. The cylinder at the deck's rear is the guard for the star-antenna base. (Patton Museum)

Viewed from above the turret opening, the Maybach HL230 P30 V-12 gasoline engine has been installed in the center compartment of a Panther Ausf. A, and the water coolers and fans have been emplaced in the upper parts of the side bays. (Patton Museum)

A Panzerbefehlswagen Panther Ausf. A turret before installation shows the turret floor, supports, and other structures. Zimmerit is on the mantlet and turret sides and front. The ring mount for a machine gun on the cupola was instituted on 1 August 1943. (Patton Museum)

A Panther Ausf. A from an unidentified unit negotiates a very rough piece of terrain in a clearing in Italy in 1944. The tube on the sponson for storing the bore-cleaning staff for the main gun is well battered, being made of relatively thin sheet metal. (Bundesarchiv)

This Panther A, tactical number 324, is evidently the same one in the preceding photo and is seen as it maneuvers over difficult terrain. The vehicle has Zimmerit, and two storage boxes have been mounted on short legs on the rear of the engine deck. (Bundesarchiv)

The white-outlined "R02" on the armor skirt of this Panther Ausf. A was the tactical number for a regimental executive officer in the unit, in this case, Hauptsturmführer Hans Gruhle of Leibstandarte SS "Adolf Hitler." The color chips to the side of the profile depict the three colors of the camouflage, top to bottom: *Olivgrün* (Olive Green: RAL 6003), *Dunkelgelb* ( Dark Yellow: RAL 7028), and *Rotbraun* (Red-Brown: RAL 8017).

The Panther in the foreground is attempting to recover the bogged Panther in the background. The closer vehicle has M.A.N.-type Zimmerit of vertical ridges in grids. Strapped above the sponson is a section of log, probably for unditching. (NARA)

Willy Langkeit, commander of Panzer Regiment "Großdeutschland" in the cupola of this Panzerbefehlswagen Panther Ausf. A confers with Gen. Hasso von Manteuffel, standing in a Schwimmwagen. (Patton Museum)

A Panther Ausf. A with the tactical number 221 on the turret maneuvers on a street in a village. The tank has the rough-textured Zimmerit associated with Daimler-Benz production and exhibits features appropriate to an Ausf. A completed up to November 1943, including the loader's periscope shield to the right of the cupola on the turret roof. The skirt panels have varying camouflage patterns and paint shades on them: a likely indication that they were scavenged from other vehicles.

A kleines Kettenkraftrad passes a Panther Ausf. A, tactical number 233, poised next to a road sign for Via Casilina in Italy. This Panther has the single exhaust tailpipes and rough-textured Zimmerit divided into grids, apparently in the style practiced by Daimler-Benz. The crew evidently is not expecting imminent action, as the 7.5cm gun is in its travel lock, a muzzle cover is present, and the antiaircraft machine gun is not installed on the ring mount on the cupola.

A column of Panther tanks proceeds along a deeply rutted, muddy road. The first two vehicles appear to be Panther Ausf. A tanks with ball-mount bow machine guns. Several of the skirt panels are missing from the lead tank, revealing the triangular piece of armor at the bottom of the sponson. The third Panther in the column has a letterbox-type cover for the bow machine gun, and the cupola is not clearly visible, so this could be a very late Ausf. D or an Ausf. A produced up to November 1943.

The commander of a M.A.N.-produced Panzerbefehlswagen Panther Ausf. A looks warily to the rear as his tank advances. On the center rear of the engine deck are the star antenna and its armored base, features associated with this vehicle. (Patton Museum)

Panthers maintain a healthy space between them as they advance across a field. The nearest tank, an Ausf. A, features Zimmerit on the skirts as well as on the chassis and turret.

A Panther Ausf. A built by Daimler-Benz between December 1943 and May 1944 advances across a plain. The rough style of Daimler-Benz Zimmerit has been applied to the vehicle. There are no pistol ports in the turret; these were deleted in December 1943.

Infantrymen, including a sniper next to the turret with a rifle with a scope, ride a Panther Ausf. A into battle. A criss-cross camouflage pattern has been painted over the base color of the tank, through which a tactical number, 625, is faintly visible.

A Panther Ausf. A with Daimler-Benz Zimmerit veers around the Arc de Triomphe in Paris in the summer of 1944. This vehicle, produced between April and June 1944, has the two cooling pipes next to the left exhaust and a jack stored between the exhausts.

Two Panther Ausf. A tanks with Daimler-Benz Zimmerit are on adjacent flatcars at a rural railroad yard. Clear details are available of the muzzle cover issued with these tanks. The tank on the left has the monocular gun sight aperture in the mantlet. (NARA)

Oberst (Colonel) Willy Langkeit, second from left, Panzer-Regiment “Großdeutschland” commander, rests with his crew by their Panther Ausf. A, tactical number 01, in Moldavia on 17 May 1944. The tank has the ball-type bow machine-gun mount. (Patton Museum)

Crewmen refuel a Panther Ausf. A, well camouflaged by pine branches in a forest. Although difficult to discern, Zimmerit in the style applied by the M.A.N. factory is present. This tank was produced between April and July 1944. (NARA)

A tow cable is attached to the rear of a Panther Ausf. A produced by M.A.N. between December 1943 and March 1944. It features M.A.N.-type Zimmerit and a jack between the exhausts and lacks pistol ports and cooling pipes next to the left exhaust. (NARA)

A M.A.N.-produced Panzerbefehlswagen Panther Ausf. A, tactical number 96, operates along a dusty road. It has the horizontally stored jack below the exhausts, a carrier for a Jerrycan next to the right exhaust, and three racks made of angle irons on the rear deck.

Crewmen arrange local camouflage around the same Panzerbefehlswagen Panther Ausf. A in the preceding photo. On the turret is an insignia featuring a black panther leaping through a circle with a vertical lightning bolt superimposed over it.

The "R02" marked on the skirts of this Panzerbefehlswagen Panther Ausf. A in Normandy indicate that it was assigned to a regimental executive officer. This vehicle was assigned to Hauptsturmführer Hans Gruhle of Leibstandarte SS "Adolf Hitler."

Crewmen decorated with Iron Crosses assemble in front of a Panther Ausf. A produced by M.A.N. between November 1943 and March 1944. It features the style of Zimmerit used by that factory, and a ball-type bow machine-gun mount is present.

In this view of a column of Panther tanks on a march, the closest tank is an Ausf. A with pistol ports on the turret, Zimmerit of the Maschinenfabrik Niedersachsen Hannover (M.N.H.) type, and the jack stored horizontally below the exhausts. A bucket and what seems to be a crate are stowed between the exhausts.

Whitewash has been applied overall to a late-production Panther Ausf. A serving with the 3rd Platoon, 8th Company, of a regiment in the 5th SS Panzer Division "Wiking" in Poland in 1944. The whitewash is relatively fresh, as the water-soluble finish tended to deteriorate when subjected to the elements.

A combat photographer crouching alongside a road next to a cargo truck catches a glimpse of a Panther Ausf. A. The action of the track guides has worn the paint off the outer rims of the inner road wheels and has polished the steel to a high sheen.

A Panther Ausf. A advances along a narrow lane. Daimler-Benz-type Zimmerit is present on the tank. Also present are the ball-type bow machine-gun mount and a single aperture on the mantlet for a monocular sight.

Horses drawing a supply cart have paused alongside a snow-plastered Panther Ausf. A while personnel in the background scan the terrain ahead. This is an M.N.H. Panther produced between September and November 1943.

During a pause in wintertime fighting, crewmen of a Panther Ausf. A take advantage of the elevation of its engine deck and the protection of the turret to make observations. This M.N.H. Panther has pistol ports and the letterbox bow machine-gun mount.

The same Panther Ausf. A shown in the preceding photograph is viewed here from the right front as Panzer personnel confer. Another Panther is in the background. The men are wearing a variety of hats and caps as well as reversible parkas with the white sides turned out.

Several bullet or shrapnel holes are visible in the skirts of this Panther Ausf. A, seen knocked out on a subtropical city street. The exhausts and the hatch cover on the rear of the turret have been blown off, and the anti-grenade screen of one of the ventilators on the left side of the engine deck is bent upward. (Patton Museum)

Two Panther Ausf. A tanks roll along a street in the city of Debrecen, Hungary, during the battle for that city in October 1944. Several of the skirt panels are missing on the lead tank; the side of the turret is covered with spare track sections.

Two Panther Ausf. A tanks of the Panzer Lehr Division were knocked out near Le Désert, France, in 1944. The Panther on the left has M.N.H. Zimmerit, and the outboard aperture for the binocular gun sight on the mantlet has been plugged. (NARA)

A member of the U.S. 4th Infantry Division peers inside the driver's hatch of a Panther Ausf. A that was knocked out by American bazookas during the fighting in Normandy on 16 July 1944. The blown-off hatch door is leaning against the left mud guard. (NARA)

A dead crewman of a knocked-out Panther Ausf. A lies under a cover in front of his tank outside of St.-Pois, France, on 5 August 1944. Grid-type Zimmerit, rather roughly scored, in the style of M.N.H. application, is visible on the glacis. (NARA)

A Panther Ausf. A knocked out by high-velocity rockets fired by a British Typhoon fighter near Bretteville, Normandy, in the summer of 1944 lies upside down on the road.

Refugees near Villedieu, France, inspect a Panther Ausf. A, tactical number 321, knocked out by U.S. gunners in the summer of 1944. Areas of the M.A.N.-style Zimmerit have spalled off the left-rear facet of the turret. (NARA)

Sometimes, tactical numbers on Panthers consisted of only two digits, indicating the number of the platoon and the vehicle. Such was the case with the Panther Ausf. A from Panzer-Regiment “Großdeutschland” in June 1944. A lightly sprayed *Olivgrün* camouflage has been applied over the *Dunkelgelb* base color. A small Stahlhelm, the symbol of this regiment, is painted on the turret to the rear of the tactical number.

No additional camouflage colors have been applied to the Panther Ausf. A of the commander of 2nd Platoon, 2nd Company, 5th SS Panzer Regiment “Wiking” in Poland in 1944. Cross-hatched Zimmerit is present on the turret and the hull.

British and American troops explore a knocked-out Panther Ausf. A, with the blown-off turret to the right. Features worthy of notice include the pistol port on the rear of the turret with a large number 2 next to it and the Demag-style ridged Zimmerit. (Library and Archives of Canada)

From the Normandy Campaign on, German Panzer troops dreaded the P-47 Thunderbolt fighter-bombers, which they dubbed "Jabos." Here, P-47 pilots get a rare chance to observe their handiwork close-up on a Panther Ausf. A they struck in July 1944. (NARA)

An Allied soldier examines a German overcoat left behind in the wreckage of a Panther Ausf. A. Blown-off road wheels are lying in the grass in the right foreground. (Library and Archives of Canada)

Allied forces encountered this abandoned Panther Ausf. A in a thicket at an undisclosed site. The access door for the engine compartment is open. A grid-type Zimmerit pattern is visible, and the left exhaust is flanked by cooling pipes for the left manifold. (Library and Archives of Canada)

Allied armored troops conduct a postmortem on a Panther Ausf. A. Though the rear of the hull was torn up, it had the early Ausf. A arrangement, with two exhaust pipes, brackets for a horizontally stowed jack below the exhausts, and no tow coupling. (Library and Archives of Canada)

A bogged or knocked-out Panther Ausf. A is under scrutiny by Allied armored personnel. Whenever possible, Allied troops would rifle a captured Panther tank for any items of intelligence value, and this highly feared and respected tank always was an object of curiosity. (Library and Archives of Canada)

Allied tank troops pose for the camera on the same Panther Ausf. A shown in the preceding photo. This was a M.A.N.-manufactured, early-Ausf. A vehicle with a cast cupola, letterbox bow machine-gun cover, and binocular main-gun sight in the mantlet. (Library and Archives of Canada)

A destroyed Panther Ausf. A lies among the ruins in front of the cathedral at Cologne, Germany, on 7 March 1945. On the previous day, from a position in the square before the cathedral, this tank had engaged in fierce duels with several U.S. tank units, which included Sherman tanks and one of the new T26E3 Pershing tanks. It was finally the Pershing that knocked out the Panther.

The Panther Ausf. A knocked out in the cathedral square in Cologne on 6 March 1945 is viewed from a slightly different angle. Thanks to several U.S. Army cameramen in the vicinity on 6 March, this tank duel was extremely well documented photographically with both movie and still cameras. The turret of the Panther was turned to the right side subsequent to the battle. (NARA)

# Panther Ausf. A

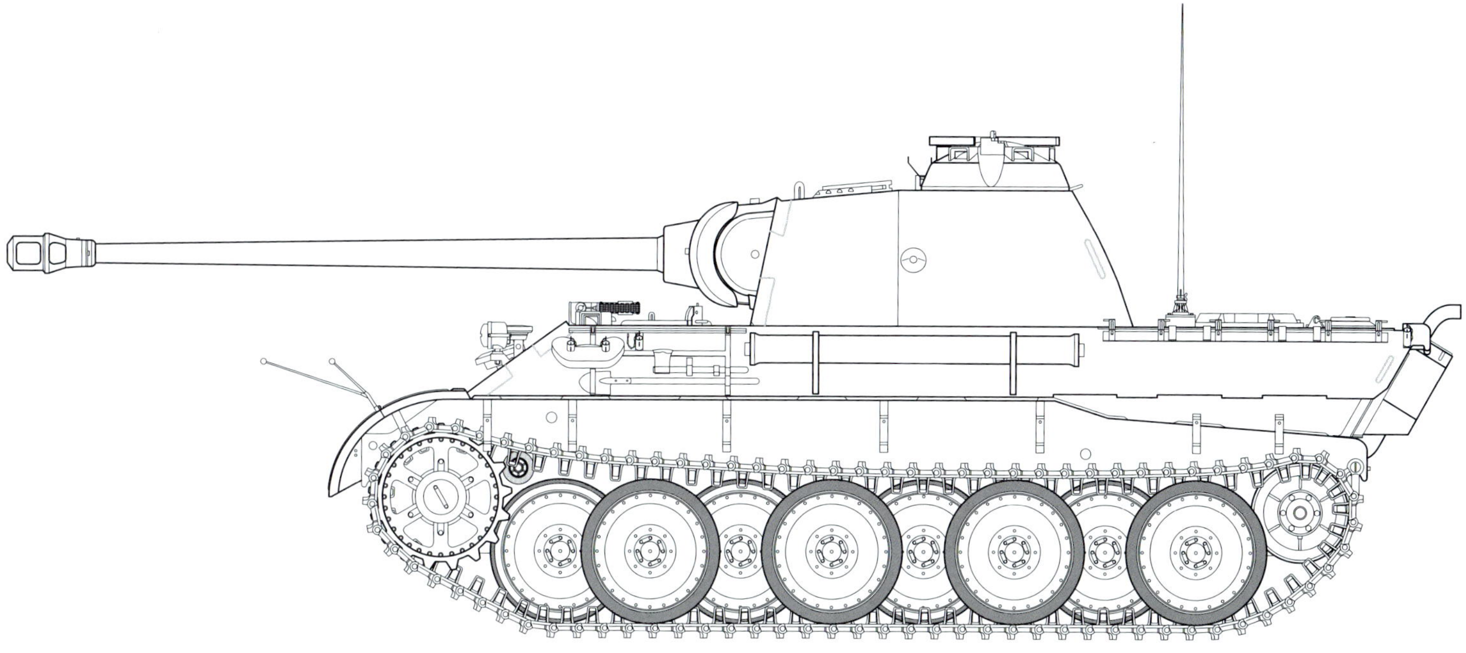

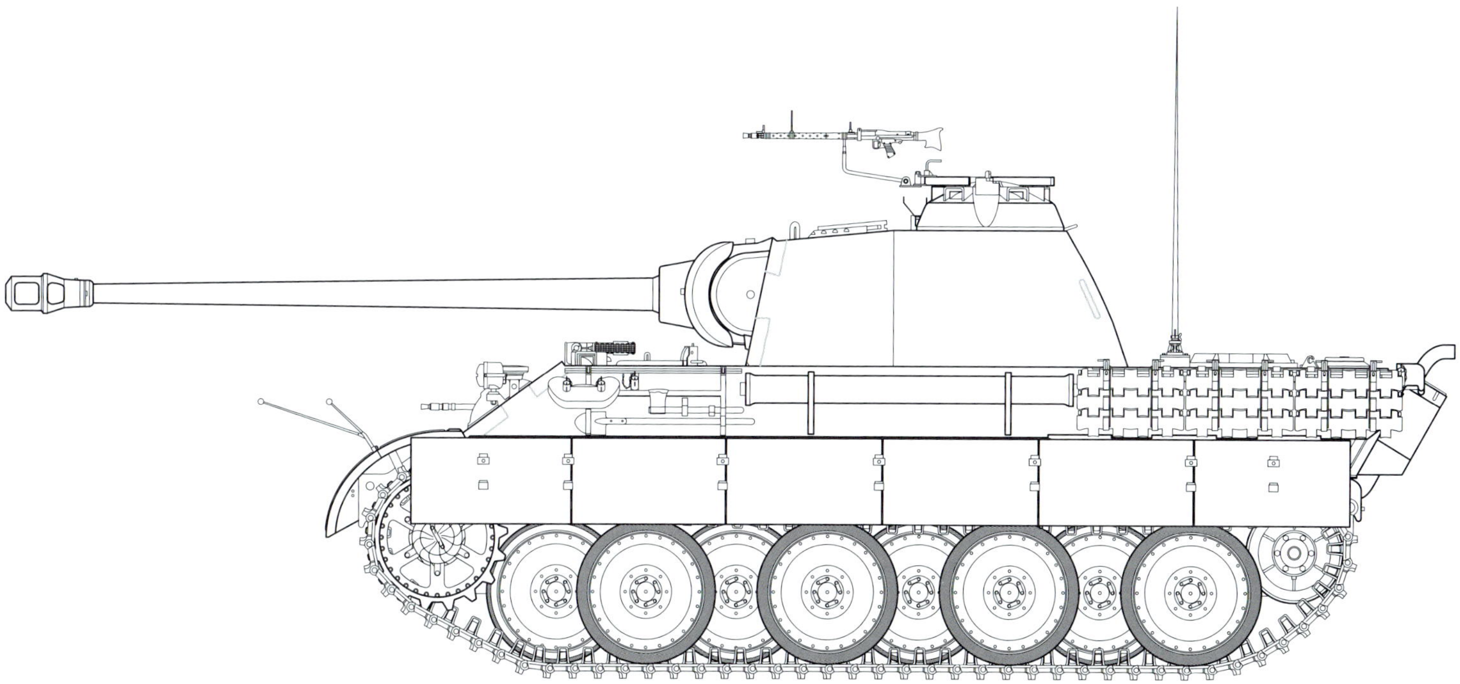

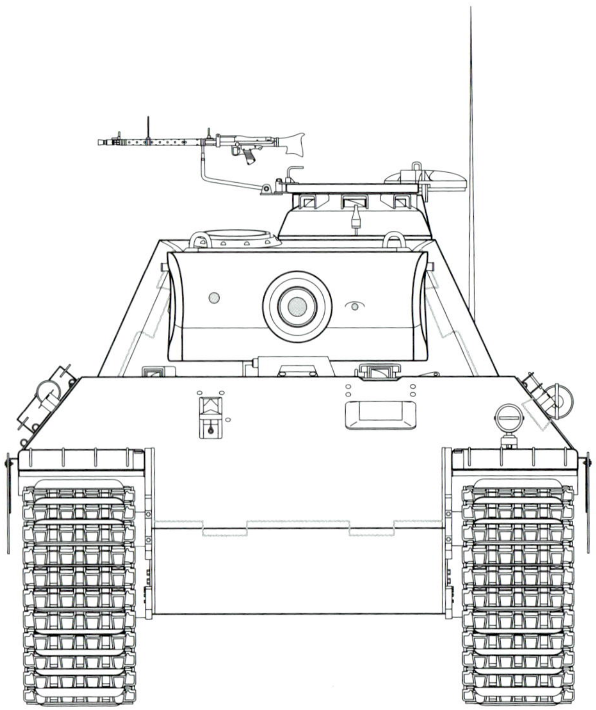

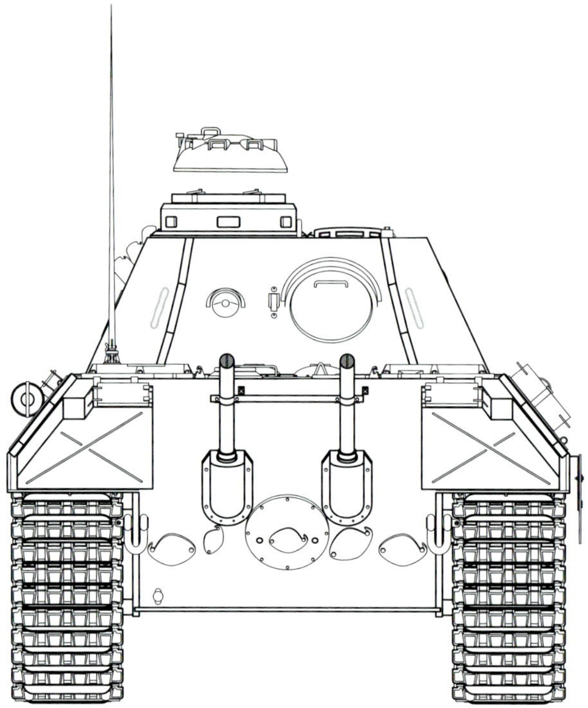

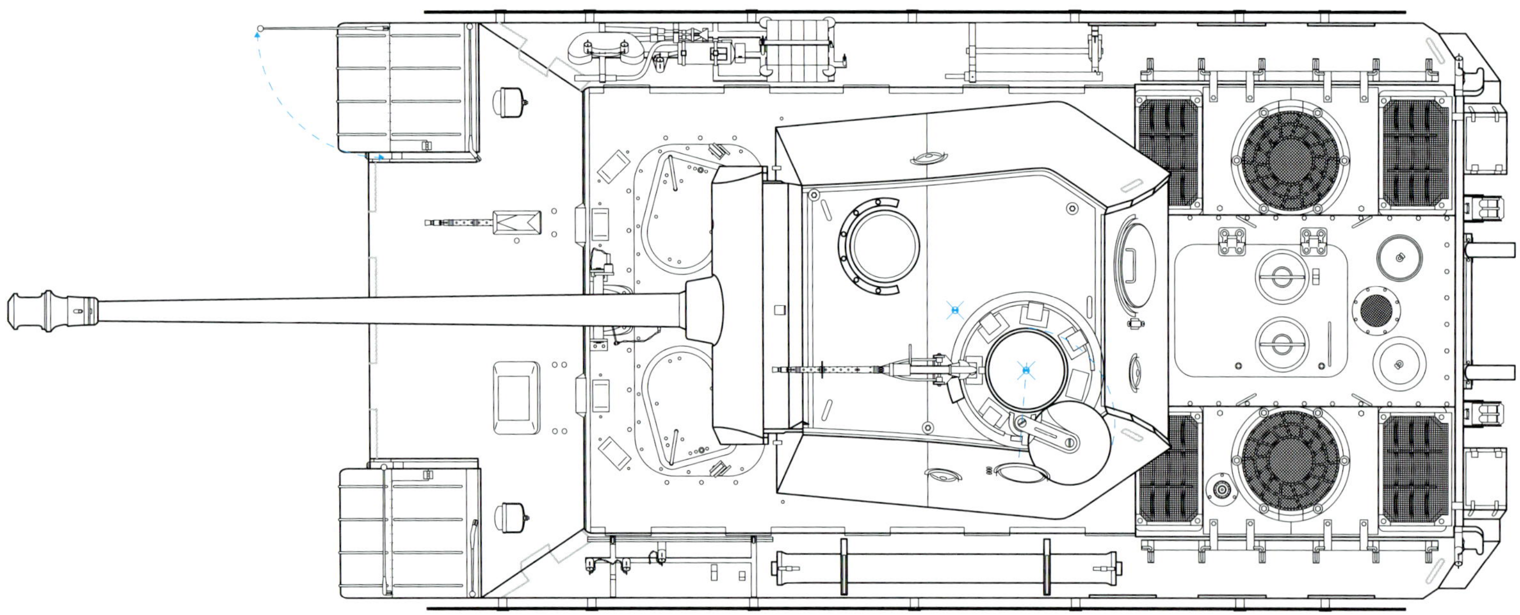

The Panther Ausf. G retained the same turret as the Ausf. A but had a newly designed hull in which the wedge-shaped armor at the bottom rear of the each sponson and the related stepped floor of the sponsons were replaced by the straight bottom edge for the sponsons and a straight sponson floor. Seen here is Panther Ausf. G chassis number 121052, completed by M.A.N. on or near 22 September 1944, featuring steel-tired 800mm road wheels, no Zimmerit, and an early example of an installation of *Kampfraumheizung* (crew-compartment heating), evidenced by the raised object on the engine deck. (Patton Museum)

Panther crewmen adjust local camouflage or enjoy a break in a courtyard on the Western Front in 1944. The nearer vehicle, an Ausf. G, has a debris guard over the mantlet, a modification instituted in August 1944, and appears to have M.A.N.-style Zimmerit.

As the war wore on, the armor protection of the Panther was found to be inadequate, while the vehicle's drivetrain was being taxed heavily, owing to the increase of the vehicle's weight well beyond original specifications. At the same time, the air war against Germany's industrial base was becoming increasingly effective. Thus, as early as 1943 design work began on the Panther II, which was to be a heavier-armored, easier-to-produce tank, having increased parts commonality with the Tiger.

Ultimately, aside from one prototype, the Panther II did not see production, but the lessons learned were incorporated into a new Panther model: the Ausf. G. The side armor of the Ausf. G was increased to 50mm, while armor in other areas, including the bottom and lower front hull, was reduced slightly to both limit the number of armor plate thicknesses needed in manufacturing as well as offset the weight gain caused by the thicker side armor. Other changes further simplified the Ausf. G's production, which began in March 1944 at M.A.N., with Daimler-Benz and M.N.H. transitioning later.

Many modern students of WWII armor argue that the Panther was the best medium tank fielded during the war. While a few Panthers did go on to serve in post-WWII armies, notably the 50 Panthers of the French Army's 503e Régiment de Chars de Combat and the 13 vehicles of Romania's 1st Armored Brigade, it is worth pointing out that the British army captured the entire Maschinenfabrik Niedersachen Hannover (M.N.H.) factory, albeit the building was damaged. The Royal Electrical and Mechanical Engineers went to considerable effort to relocate the tools and equipment needed for production to Laatzen, south of Hannover, before resuming production with M.N.H. employees. There they made nine Panthers and 12 Jagdpanthers for study – but did not continue production for either their own use, or for sales to other governments. However, the British-built Panthers were the last of the type to be completed.

Tank crewmen and infantrymen congregate on a Panther Ausf. G in France in 1944. The rear panel of the right armored skirts is missing, revealing just below the spare tracks the thin fender on the hull to which the tops of the Ausf. G-type skirting abutted

On some Panther Ausf. Gs, such as this one with M.A.N.-style Zimmerit produced between June and August 1944, the cylindrical container for the bore-cleaning brush and staff was moved from the left side of the vehicle to the rear of the engine deck.

Panther tanks are being transported on railroad flatcars. The tank to the front of the partially visible vehicle in the foreground is a Panther Ausf. G. The exhausts have welded guards, introduced in May 1944, with a Zimmerit coating, and sheet-metal shields for the tailpipes, a June 1944 modification. The Zimmerit on the rear storage bins is the variety with vertical ridges and diagonal scratch marks.

Two crewmen of a Panther Ausf. G, heavily camouflaged with tree branches, wait expectantly as thick smoke billows up from an explosion to the front during a battle in France in 1944. This vehicle has cast-steel exhaust guards and sheet-metal tailpipe shields, both of which have Zimmerit coatings. The Zimmerit is of the rough-textured pattern associated with Daimler-Benz's application style.

A M.A.N. Panther Ausf. G produced between June and August 1944 has halted on a road in France in 1944. The spare roadwheel on the turret is an early one with 16 bolts around the perimeter. The thin fender is visible and damaged toward the front.

A Panther Ausf. G of 1st SS Panzer Division Liebstandarte SS "Adolf Hitler" approaching the Arc de Triomphe in Paris in the summer of 1944 has the rough-textured, rather amorphous Zimmerit associated with Daimler-Benz's style of application. The handle on the turret's rear hatch was introduced in June 1944. Tow cables are attached to clevises mounted on the rear of the hull.

Another Panther Ausf. G photographed at the Arc de Triomphe exhibits the style of Zimmerit applied at the M.A.N. factory and the type of welded exhaust guards that came into production in June 1944. Several of the right skirt brackets are visible. On the round access plate on the rear of the hull is the tow coupling for attaching tow bars: a feature introduced with Ausf. A.

Two Panzer crewmen by a tent and another one by the rear of a Panther Ausf. G take a break. The tactical number marked on the side of the turret is "I02," which stands for the vehicle of the assistant commander of the 1st Battalion. The full extent of the thin fender running along the side of the sponson is visible. Brackets along the underside of the fenders served to hold the skirts in place.

Two crewmen are engaged in a repair or replacement operation on the left track of a M.A.N.-built Panther Ausf. G. Details of the Kgs 64/660/150 track are visible. The number 660 referred to the width of the track links, 660mm. The number 150 represented the pitch, or the front-to-rear dimension of each link, in millimeters. There were 87 track links per side. These tracks were a single-dry-pin type.

Infantrymen take advantage of a Panther Ausf. G as a battle taxi. The vehicle exhibits M.A.N.-pattern Zimmerit and the style of welded exhaust guards introduced in May 1944 and the sheet-metal tailpipe shields instituted in June 1944.

This Panther Ausf. G built in mid-September 1944 was equipped with an infrared searchlight and scope on the cupola. Particular to Daimler-Benz was the "ambush" camouflage, which featured small, contrasting splotches of paint on the *Dunkelgelb, Rotbraun,* and *Olivgrün.* (Thomas Anderson)

Canadian troops fired two PIAT (Projector, Infantry, Antitank) rounds into this Panther Ausf. G, knocking it out in Bretteville, Normandy, on 20 June 1944. The tank, with M.A.N.-type Zimmerit, has features consistent with April-June 1944 production. (Library and Archives of Canada)

On this knocked-out Panther Ausf. G in Humain, Belgium, on 28 December 1944, jutting above the engine deck is a crew-compartment heater, an accessory that was installed on Panther Ausf. G assembly lines starting in October 1944. (NARA)

This 2nd SS Panzer Division Panther Ausf. G was disabled and froze fast to the Many-Grandmenil road near Houffalize, Belgium. U.S. engineers used a bulldozer to move the tank, which toppled over with soil firmly frozen to its now upside down tracks. (NARA)

The turret of a Panther Ausf. G bears the tactical number R01, indicating the tank of the commander of a Panzer regiment. This vehicle features two distinct camouflage schemes, including an "ambush" camo used by M.A.N. on the skirts. (Patton Museum)

Three-color camouflage is present on a Panther Ausf. G with markings for the eighth vehicle, 2nd Platoon, 1st Company, SS Panzer Regiment 12 in the 12th SS Panzer Division "Hitlerjugend" in Normandy in 1944. The early-type mantlet lacks the chin that later was incorporated into the mantlet.

A Panther Ausf. G with the late mantlet with a chin is painted in the so-called "ambush" camouflage introduced in the final months of World War II. This camouflage consisted of *Dunkelgelb, Olivgrün,* and *Rotbraun,* with small splotches of *Dunkelgelb* applied here and there on the two darker colors.

A distinctive three-color camouflage has been applied to this Panther Ausf. G, with *Olivgrün* areas with hard-edged *Rotbraun* borders applied over the *Dunkelgelb* base color on the main gun, the turret, and the upper hull. No tactical numbers or unit markings are present.

A Panther Ausf. G with M.A.N.-type Zimmerit is stuck in a roadside ditch in France on 16 August 1944. This tank was built before July 1944, as it lacks the three *Pilze* (sockets) for installing a boom on the turret roof. These sockets came into production in June 1944.

An M26 tractor manned by the 486th Ordnance Evacuation Company, 84th Infantry Division, Ninth U.S. Army, hauls a captured, trackless Panther Ausf. G through Geilenkirchen, Germany, on 4 December 1944. The tank has M.A.N.-pattern Zimmerit. (NARA)

The same trackless Panther Ausf. G seen in the preceding photo is shown from a different perspective in the middle background as an M26 tractor tows it along a street in Geilenkirchen. American forces had knocked out this tank in the outskirts of that town. (NARA)

Members of the U.S. 2nd Armored Division investigate a snow-caked Panther Ausf. G on the outskirts of Grandmenil, Belgium, on 4 January 1945. The crew of the Panther abandoned it when a cartridge became jammed in the 7.5cm gun. (NARA)

A 9th Armored Engineer Battalion Caterpillar Model D7 tractor with a bulldozer pushes an abandoned, snow-plastered Panther Ausf. G out of the way along a road near Manhay, Belgium, in early January 1945. (NARA)

A U.S. Army Sherman tank of the 2nd Armored Division rolls past a knocked-out Panther Ausf. G outside of Grandmenil, Belgium, on or about 2 January 1945. Over the driver's periscope is a sheet-metal rain guard, a feature introduced in August 1944. (NARA)

An abandoned Panther Ausf. G of 9th SS Panzer Division sits in a street in Sterpigny, Belgium, on 20 January 1945. The vehicle has an overall dark appearance, with patches of snow on it. Toward the rear of the side of the turret is the tactical number 412. (NARA)

A British or Commonwealth soldier contemplates the chewed-up front mudguard of a Panther Ausf. G in a town in Europe. What appears to be M.A.N.-type Zimmerit is visible. The running gear is destroyed. On the mantlet is a monocular sight aperture. (Library and Archives of Canada)

An Allied soldier looks over an abandoned M.A.N.-built Panther Ausf. G, tactical number 132, in Normandy in 1944. One of the armor skirt panels was blown off and is leaning on the front of the turret. The vehicle has characteristics of April-to-June 1944 construction. (Library and Archives of Canada)

British or Commonwealth personnel inspect a captured Panther Ausf. G. This tank has the vertically stored 20-ton jack, introduced to production in February 1944, and the tow coupling on the round access hatch on the rear of the hull, introduced in April 1944. (Library and Archives of Canada)

Another view of the tank above as Allied troops take stock of a knocked-out Panther Ausf. G, tactical number 135. Some of the troops are reading documents that likely they have discovered in the tank. The Zimmerit on the vehicle and the placement of the Balkenkreuz are M.A.N-style. (Library and Archives of Canada)

# Panther Ausf. G

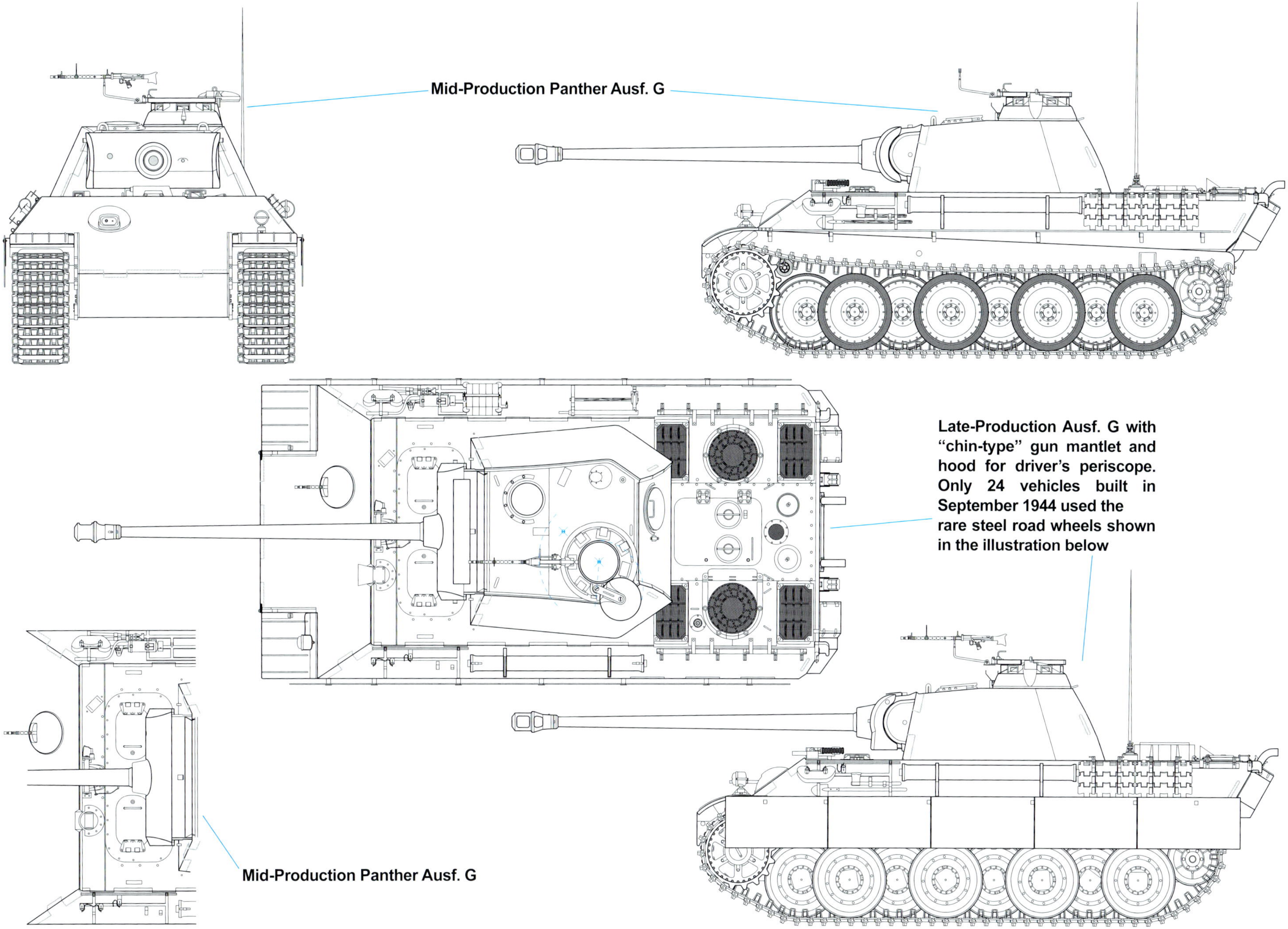

British or Commonwealth crewmen are staffing this captured and operational M.A.N. Panther Ausf. G in Europe in late 1944 or early 1945. This vehicle has a debris guard over the mantlet and a standard mantlet without the "chin" introduced in September 1944. (Library and Archives of Canada)

A Sherman ARV tows a captured Panther Ausf. G in the winter of 1944-1945. The 7.5cm gun and mantlet have been removed from the turret. The Balkenkreuz on the front of the sponson is situated in accordance with M.A.N. factory painting practices. (Library and Archives of Canada)

A G.I. looks over a knocked-out Panther Ausf. G at Kelberg, Germany, on 11 March 1945. The presence of a crew heater on the engine deck and a ring mount on the cupola indicate that the manufacture of the vehicle was completed between October 1944 and January 1945. (NARA)

The tactical number 111 is faintly visible through the soot on the turret of this Panther Ausf. G. When it was knocked out, the flames caused most of the rubber tires on the road wheels to burn. A "chin" mantlet with extended bottom to deflect shots is present. (NARA)

A Free French soldier prepares to spray a new coat of paint on a captured Panther Ausf. G in a city in France on 18 September 1944. The tank was being prepared for dispatch to Paris as a war trophy. Faintly visible on the roof of the turret in line with the center of the cupola from this angle is one of three *Pilze* sockets welded to the roof to hold a jib boom for heavy-lifting and repair operations in the field. These sockets began finding their way onto Ausf. G turrets in the factories in June 1944. The prominent ledge above the mantlet is the front edge of the debris guard, introduced to production in August 1944. (NARA)

A U.S. Army cargo truck has just passed a Panther Ausf. G knocked out by elements of the 2nd Armored Division near Brissonville, Belgium, on 29 December 1944. This Panther features a crew-compartment heater and welded guards for the exhaust pipes. (NARA)

A Panther Ausf. G, tactical number 252, lies abandoned on a battlefield, its right mudguard torn up. The presence of both a debris guard over the mantlet and factory-applied Zimmerit indicate that this tank was completed in August or September 1944. (Patton Museum)

An antitank team of the 1st Infantry Division inspects a Panther Ausf. G knocked out by a bazooka near Fernegierscheid, Germany, on or around 26 March 1945. The most conspicuous damage is the torn-up sprocket. This Panther is equipped with the chin mantlet. (NARA)

Parked next to a Tiger tank is a Panther Ausf. G, tactical number 135. The placement of the Balkenkreuz at the front of the sponson is the specific practice of the M.A.N. factory, and the Zimmerit is typical M.A.N.-style, with vertical ridges set inside grids. (Patton Museum)

This Panther Ausf. G exhibits characteristics in keeping with a completion date of April to June 1944, as it lacks the *Pilze* sockets on the turret roof for a jib boom. The placement of the Balkenkreuz and the pattern of the Zimmerit are of the M.A.N. style. The Zimmerit consists of vertical ridges laid out in horizontal rows, not discernible grids. The cylinder for storing the bore-cleaning brush and staff is battered and the cap on the front end is detached. Of interest is the fact that the tactical number, 308, has been painted over a larger, former tactical number, 126. (Patton Museum)

All the Panthers were powered by either the HL210, shown here, or the HL230TRM gasoline engine. Designed by Maybach, the engine was produced by Maybach, Auto Union, and Daimler-Benz. (NARA)

With the bombed out massive Vereinigte Fränkische Schuhfabriken (United Franconian Shoe Factories) rising in the background, a trio of combat-worn Panthers await unloading from railcars at the M.A.N. plant. The rebuild for which they were sent to Nürnberg will never come. (NARA)

Besides producing new Panthers, M.A.N. also remanufactured battle-weary Panthers regardless of manufacturer. Incoming tanks were dismantled on a disassembly line, then reassembled using new or refurbished mechanical. This Panther G hull, coincidentally built by M.A.N., has been stripped of many of its internal components. (NARA)

This Daimler-Benz built Panther was among several found on railcars at the M.A.N. plant when it was captured in April 1945. Heavy bombing raids had completely disabled the plant, and even rebuilding tanks in the facility had been stymied as a result. (NARA)

# Panther G ersatz M10

During the Ardennes Offensive in December 1944, the Germans sent into battle a number of Panther Ausf. G tanks disguised as U.S. M10 tank destroyers. Here, U.S. soldiers inspect one of these vehicles that has lost its left track and become mired along a road. (Patton Museum)

The left side of the mantlet of this fake M10 tank destroyer based on a Panther Ausf. G has been shot up. Panels were applied to the Panther to replicate the angular M10, and U.S. insignia and bogus unit markings were painted on these vehicles. (Patton Museum)

The left rear of the turret of the same fake M10 is viewed close-up. The turret was traversed toward the right to about the 2 o'clock position. The rear of the Panther's turret is to the right, and the fake side plate is to the left. (Patton Museum)

The badly damaged false mantlet of the same imitation M10 seen in the preceding photo is viewed close-up. The shape of the stock Panther Ausf. G mantlet without the shot trap on the bottom is under the ripped sheet metal of the false mantlet. (Patton Museum)

The roof of the same fake M10 is seen from above, facing to the rear. To the left rear of the turret, the Panther's cupola has been removed and a flat, two part hatch fitted over the cupola's opening. A crude U.S. insignia is painted on. (Patton Museum)

The Panther Ausf. F was an experiment to mate a newly designed Schmalturm (narrow turret) to a Panther Ausf. G chassis. One of these experimental vehicles is shown in a 20 August 1944 photograph. No tools were mounted on the chassis for these experiments. (Thomas Anderson collection)

The Schmalturm had a narrow front, presenting a smaller area to frontal fire. The nearly conical mantlet was designed to deflect shots and prevent a shot trap that would permit rounds striking the bottom of the mantlet to penetrate the hull roof. (Patton Museum)

The sides of the Schmalturm feature small loops for fastening camouflage materials. The commander was provided with a cupola that was lower than the one on the Panther Ausf. G, and was equipped with seven periscopes with armored hoods. (Patton Museum)

Unlike the standard Panther commander's cupola, the hatch of which rose then swung to the side, the commander's hatch on the narrow turret was hinged at the back. At the rear of the turret is an escape hatch and a pistol port. (Patton Museum)

A curved rain deflector was attached to the front of the turret over and to the sides of the mantlet. The left lifting eye had broken off. Jutting from the upper part of the sides of the turret were guards for rangefinder objectives. (Patton Museum)

## Bergepanther

As the weight of German armored vehicles increased, so too did the challenge of recovering broken down or battle damaged Panzers. To address this, a recovery vehicle derivative of the Panther was produced. The first, and most basic, type of these recovery Panthers – or Bergepanthers – were simply turretless Panther Ausf. D chassis with a cover fitted over the turret opening, completed by M.A.N. in mid-1943.

There followed 30 vehicles built by Henschel and based on the Panther Ausf. A chassis with a heavy towing bracket and provisions for an unditching beam, but little further recovery gear. Forty more units followed that added a superstructure with folding wooden sides to the equipment installed on their predecessors.

In time the design progressed and the Bergepanther began to be fitted with a 40-ton capacity recovery winch fitted in the area normally defined as the fighting compartment. A jib boom for material handling was provided and a hinged ground anchor (spade) was installed on the rear of the vehicle. The latter was to prevent the winch from merely dragging the Bergepanther backwards during heavy recovery operations.

In October 1944 the Bergepanther began to be based on the Panther Ausf. G as that model replaced the Ausf. A in tank production. Ultimately, M.A.N. built only the 12 original, very basic Bergepanthers, Henschel built 70 Bergepanther Ausf. A, and Daimler-Benz a further 40. Demag built 123 Bergepanther Ausf. A as well as 94 Bergepanther Ausf. G, even though Demag never built any Panther Ausf. G combat tanks.

**Seen in Nürnberg in December 1943 is the first fully-equipped Panther recovery vehicle. The lifting boom was specific to the Bergepanther Ausf. A. (Walter Spielberger)**

**Winch-equipped Versuchs (test) Bergepanther Ausf. A chassis number 212161, assembled by Henschel in September 1943, features the early model of the *Sporn* (spade). This spade design was susceptible to digging into the ground too deeply. To prevent this situation, a new spade style was introduced on production vehicles, one having two steel plates at the top of the blade to limit the depth to which it could dig in, as well as longer arms. This spade would be installed on December 1943-January 1944 vehicles. Above the glacis is an armored shield for a 2cm cannon. (Walter Spielberger)**

**Henschel Bergepanther Ausf. A chassis number 212177, captured by the Soviets, is on display in Moscow, with two identifying placards in Russian attached to the mudguards. It features the Ausf. A's driver's visor and machine-gun port on the glacis, and two handholds. (Patton Museum)**

The upper part of the glacis of the captured Bergepanther Ausf. A 212177 is shown close-up, allowing a clear view of the support structure for a 2cm automatic cannon and armored shield. At the upper front of the support is a socket for the gun mount. (Patton Museum)

The support for the 2cm cannon on a Bergepanther Ausf. A is seen from the right rear, with the glacis below it. Three triangular gussets are welded to the rear of the socket for the gun, for reinforcement, and three large gussets are welded to the rear of the support. (Patton Museum)

In order to allow freer play for the winch cable, this spade replaced the early version and featured cutouts for the cable as well as a reinforced design and longer arms. The exhaust tailpipes were extended to clear the raised spade; between them is a winch cable roller. (Patton Museum)

The rear of Henschel Bergepanther Ausf. A 212177 is shown while the vehicle was being exhibited in Moscow. Several access plates are removed or open. At the bottom of the rear plate of the hull are brackets for the spade and, at the center, a tow bracket. (Patton Museum)

The exhausts of Bergepanther Ausf. A 212177 have horizontal dividers in the openings, typical for a Panther-based vehicle. Between the exhausts is the winch-cable guide, fitted with two vertical rollers at the rear (seen here) and two horizontal rollers to the front. (Patton Museum)

Two steel pads for pushing disabled vehicles were mounted on brackets on the bow of the Bergepanther. Henschel began installing these pads on the Bergepanther, starting with chassis number 212161, and they continued to be a feature on the Bergepanther Ausf. G. (Patton Museum)

The winch-cable guide on Bergepanther Ausf. A chassis number 212177 is seen from the front, showing the two vertical rollers at the front of the guide. This guide allowed the winch cable to be paid out or pulled in without chafing or getting caught on the hull. (Patton Museum)

Pushing pads on the Bergepanther Ausf. A feature five conical bolt heads, as seen in this view of the right pushing pad on Bergepanther Ausf. A chassis number 212177. To the side of it is the right-front tow shackle, secured with a heavy pin to a perforated fitting built into the side plate of the hull. (Patton Museum)

This Bergepanther Ausf. A appears to have been assembled in September 1944, utilizing a hull produced in 1943. Zimmerit is present on the lower part of the hull, but absent on the upper part of the hull, probably a result of the conversion process. The Germans discontinued the application of Zimmerit on armored vehicles in September 1944, so the supposition is that the conversion of the vehicle spanned the cutoff date for Zimmerit application. The boom is installed, and a sheave is suspended from it by a chain. A wooden beam is stored on the side of the superstructure. (Patton Museum)

The right pushing pad on the bow of the same Bergepanther Ausf. A seen in the preceding photo is missing. The tall steel plates on the periscope guards originally were for holding an antiaircraft machine-gun mount but also served to support the foul-weather cover. (Patton Museum)

The same Bergpanther Ausf. A completed in or around September 1944 is viewed from the rear. The spade was the late type that featured two rectangular steel plates on the top of the blade to keep it from sinking too far into the ground when deployed. (Patton Museum)

The towing bracket on the rear of the September 1944 Bergepanther Ausf. A was the type relocated from the bottom of the rear plate of the hull to the round access plate higher up on that plate. This location was found to be more effective for the towing fitting. (Patton Museum)

The late-style spade is lowered on the same Bergepanther Ausf. A. The tailpipes of the exhausts are the extended variety, to clear the spade when in the travel position. Each of the two extension plates at the top of the spade has three braces for extra stiffening. (Patton Museum)

The manufacturing firm M.A.N. began development of a 40-ton *Seilwinde* (winch) for the Bergepanther in mid-1943. The winch took a year to develop, and it was installed in the body below the superstructure, as seen in this photo taken from overhead. (Patton Museum)

The winch cable exited the rear of the superstructure through a round port. Part of the 40-ton winch is visible to the lower right, and the engine deck of the Bergepanther is to the upper left. A cable guide with a vertical roller is to the inside of the port. (Patton Museum)

The rear cable guide of a Bergepanther is viewed from the front. The winch cable was attached to the spade in order to allow the winch to raise and lower it, and the cable also was used to recover disabled or mired vehicles. (Patton Museum)

These two sheaves were part of the standard accessories of the boom. They were deployed in various ways to multiply the pulling force of the boom cable. (Patton Museum)

British-captured Bergepanther Ausf. A chassis number 175644 undergoes evaluation at the School of Tank Technology in England. Demag completed this vehicle in September 1944, and the presence of Zimmerit only on the lower hull is consistent with that date. (Patton Museum)

The spade of Bergepanther Ausf. A chassis number 175644 is dug in firmly, and the winch cable is taut and under load during evaluations at the School of Tank Technology. The round hole in the rear of the superstructure for the winch cable is visible. (Patton Museum)

The crew of this Bergepanther abandoned it on a street in a village in France in 1944. The wooden sideboards of the top of the superstructure are missing, and no spade is installed. Some of the diagonally crosshatched Zimmerit has been knocked off of the glacis. (Patton Museum)

A late-type spade is mounted on this Bergepanther, and the wooden extensions have been removed from the top of the superstructure. On the side of the upper hull are two heavy-duty, C-shaped tow hooks. Zimmerit is present on the upper and the lower hull. (Patton Museum)

British troops evaluate a captured Bergepanther Ausf. A. Only the front and the rear sideboards are installed on the superstructure. On the hull next to the superstructure are a roll of wire and a wooden block to support the jack when lifting the vehicle. (Patton Museum)

A Bergepanther is parked at a vehicle-collection point. The vehicle has a prominent, rough Zimmerit application, and the engine-access hatch on the rear deck is open. The left sideboard of the superstructure, called a *Ladebrücke,* is in the folded-down position. (Patton Museum)

Abandoned at Morhet, Belgium, six miles from Bastogne, this snow-covered Bergepanther is seen on 27 January 1945. U.S. Third Army Ordnance Intelligence intended to send the vehicle back to the United States for further analysis. (NARA)

The Bergepanther Ausf. A captured in Morhet was almost new. It had been completed by Demag in October 1944, issued to Pz. Lehr.Div. on 7 December 1944, and was captured that same month. In 1945 it was shipped to Aberdeen for evaluation. (NARA)

A Bergepanther with a late-style spade is parked at a depot among British armored vehicles. A well-maintained camouflage scheme constituting at least two colors is present on the vehicle, including the spade, suggesting either a repaint job or a barely-used vehicle. The extended tailpipes are visible between the extension plates on top of the spade blade. Visible on the left arm of the spade is a reinforcing plate with a lightening hole, of the pattern typically seen on Bergepanthers produced by Demag in the final months of 1944. (Patton Museum)

The Jagdpanther tank destoyer mated the Panther tank chassis, drive train, and hull with a special, armored superstructure and the powerful 8.8cm Panzerabwehrkanone PaK 43/3 L/71 gun. Mühlenbau und Industrie A.-G. (MIAG) of Braunschweig, Germany, was sole maker of the vehicle through October 1944, after which the firms M.N.H. and Maschinenbau und Bahnbedarf A.-G. (M.B.A.) produced a small number of Jagdpanthers. This is one of the first production MIAG Jagdpanthers from early 1944. (Thomas Anderson)

The Jagdpanther, or Hunting Panther, combined the automotive components of the Panther with the powerful 8.8cm PaK L/71 gun used on the Tiger II, mounted in a relatively low-profile tank-destroyer hull. The Jagdpanther was intended to be used as an offensive weapon by heavy tank destroyer battalions. The lack of a rotating turret meant that the Jagdpanther itself had to be protected by infantry and/or tanks.

A full-sized wooden mockup of the vehicle that would become the Jagdpanther was displayed by Krupp in November 1942, even though a month previous the decision had been made to transfer the project to Daimler-Benz. In December 1942 the project was passed off again, this time to Mühlenbau und Industrie A.-G. (MIAG) in Braunschweig. It would be almost a year before MIAG finalized the design and completed two prototype vehicles.

The first five production vehicles were completed by MIAG in January 1944 with the goal being to raise output to 150 vehicles per month. That was not to be, however, as Allied bombers began to target MIAG's Braunschweig plant, delaying production. To offset this, contracts were issed for additional Jagdpanther production by Maschinenfabrik Niedersachsen-Hannover (M.N.H.) in Hannover as well as Maschinenbau und Bahnbedarf A.-G. Nordhausen (M.B.A.) at Potsdam-Drewitz. It was hoped that deliveries from the two additional firms would begin in November.

Allied bombing, not only of assembly plants but also infrastructure and supplier plants, severely limited Jagdpanther production. While the chassis numbers of the M.B.A.-produced vehicles are not known, it is known that only 33 Jagdpanthers were completed by the firm. M.N.H. completed 112 vehicles, chassis number 303001 through 303112, and MIAG completed 268 units, chassis numbers 300001 through 300268.

**Only the driver's right periscope is installed in the Jagdpanther, but the embrasure for the left periscope is visible next to it. A C-hook, tow cable, and tubular storage container for a bore-cleaning brush are on the vehicle's left side above the skirts. (Patton Museum)**

**At first glance, these troops sitting on an early Jagdpanther with dual vision ports look like Soviet tankers, but the soft helmets several of the men are wearing are not the Soviet type, and a close inspection of the photo reveals that they are wearing German Panzer crew uniforms. Three fore-and-aft stiffeners with rounded tops are stamped into each of the front mudguards. (Thomas Anderson collection)**

Installing the 8.8cm PaK 43/3 L/71 gun in the Jagdpanther necessitated mounting it in a deep socket or recess built into the glacis. This socket formed a natural shot trap, providing an easy way for shells striking it to penetrate the glacis. To counter that vulnerability, it was necessary to provide the gun with as effective a mantlet as possible. After much designing and redesigning, this mantlet was settled on, although it did not totally cover the socket in which it was installed.

Jagdpanthers produced in June 1944 retained only the driver's right periscope, with the opening for the left periscope covered and a straight, angled rain guard replacing the peaked rain guard. This guard was deleted from Jagdpanthers produced from July 1944.

## Jagdpanther Specifications

| | |
|---|---|
| Length | 9.88 meters |
| Width | 3.45 meters |
| Height | 2.72 meters |
| Weight | 46 metric tons |
| Fuel Capacity | 700 liters |
| Maximum Speed | 46km/hr |
| Range, On Road / Cross Country | 160km / 80km |
| Crew | 5 |
| Communications | Fu 5 and Fu 2, Intercom |
| Weapon, Main | 8.8cm Panzerabwehrkanone PaK 43/3 L71 |
| Weapon, Secondary | 7.92mm MG 42 |
| Ammo Stowage, Main | 60 rounds |
| Ammo Stowage, Secondary | 1,200 rounds |
| Engine Make/Configuration | Maybach V-12 |
| Engine Model | HL230 TRM P30 |
| Engine Displacement | 23.1 Liters |
| Engine Horsepower | 600 @ 2,500 R.P.M. |

This Jagdpanther, chassis number 300054, assigned to Major Erich Sattler, 3rd Regiment, Schwere Panzerjäger Abteilung 559, was knocked out near Hechtel, Belgium, in September 1944. This was a Befehls-Jagdpanther: a command vehicle with extra radio equipment in the left rear of the superstructure. Four projectiles had pierced the right side of the vehicle: one toward the rear, and three closely-grouped ones below the rear of the superstructure. This Befehls-Jagdpanther now is in the collections of the Imperial War Museum. (Patton Museum)

The Befehls-Jagdpanther captured at Hechtel and now in the Imperial War Museum is viewed head-on. The Jagdpanthers were equipped with the type Kgs 64/660/150 tracks. Each link had six "chevrons": raised, angled cleats that were incorporated into the treads. (Patton Museum)

On the Befehls-Jagdpanther captured at Hechtel, Belgium, painted in small characters on the lower lip of the gun socket is "DTD 3081." The right fender, made of sheet metal, had been severely damaged. The 8.8cm gun had the early, monobloc barrel. (Patton Museum)

The front skirt panel on the right side is missing on this Jagdpanther of *schwere Panzerjäger Abteilung 654* during training at Mailly-le-Camp, France. The 8.8cm gun has the sectional, two-piece gun barrel rather than the earlier single-piece monobloc gun barrel. (Patton Museum)

In another view of the same Jagdpanther, the commander is in the hatch on the right side of the roof. To his front is a rotating plate with a periscope and guard, to the front of which is, protruding through an opening in the plate, a scissors telescope. (Patton Museum)

The Befehls-Jagdpanther captured at Hechtel in September 1944 is viewed from above. The vehicle had cross-hatched Zimmerit and a single driver's periscope with no rain guard: features consistent with a Jagdpanther produced in July or August 1944. The doors are missing from the commander's and loader's hatches on the superstructure roof. Toward the left front of the superstructure roof is the sliding shield for the gunner's periscope. (Patton Museum)

Members of the 1st Antitank Regiment, Royal Canadian Army, inspect a knocked-out Jagdpanther in the Reichswald on 14 March 1945. The soldier squatting to the left is pointing to the right final drive, from which the sprocket has apparently been blown off. (Library and Archives of Canada)

A British or Commonwealth soldier is standing on a captured Jagdpanther at a vehicle park. The Jagdpanther is the final model, with a single visor for the driver and the bolted-on gun mount. The fenders and mud guards are damaged, and the headlight cover is detached. To the left is a Panzer IV/70. (Library and Archives of Canada)

The same knocked-out Jagdpanther depicted in the preceding photograph is viewed from the front left. The 88mm round that is lying in the foreground in the preceding photo has been propped up, providing a sense of the impressive scale of the ammunition. The tank's jack is lying in the left foreground. (Library and Archives of Canada)

One of the first eight production Jagdpanthers, this vehicle assigned to schwere Panzerjäger Abteilung 654 crosses a plain during training exercises at Mailly-le-Camp, France, in early August 1944. The Zimmerit extended over the Kugelblende but not to the socket for the 8.8cm gun or between the driver's periscope and rain guard.

A Kugelblende (ball machine gun mount) was located to the right of the 8.8cm gun on the glacis of the Jagdpanther. The machine gun, the 7.92mm MG 34, is not installed here. The crosshatch-pattern Zimmerit extended to the mud guards. (Patton Museum)

A Jagdpanther of June 1944 construction is positioned in a pine forest. A shovel is stored on the left side of the superstructure below a wire cutter, a C-hook, and a tow cable. The lower front corner of the front skirt panel has been bent back.

Two crewmen go about their duties to the front of and on top of the Jagdpanther with the bent-back corner of the front skirt panel. The typical crosshatched Zimmerit pattern is evident on the hull and the superstructure.

A crewman uses a spray gun to paint a camouflage pattern on the right rear of a Jagdpanther of *schwere Panzerjäger Abteilung 654.* Worthy of notice is the absence of Zimmerit on the side of the sponson where the spare track links were stored.

A Jagdpanther sits between a road and a plowed field. The left mudguard, a sheet-metal component that lacked much rigidity, is severely damaged. The storage tube for the bore-cleaning rod has been moved from the superstructure to the rear of the engine deck. (Patton Museum)

The Befehls-Jagdpanther in the preceding photograph is viewed from a different perspective. This vehicle was manufactured by MIAG and was assigned chassis number 300054. The rear of the right track is severed, and the on-vehicle tools have been stripped. (Patton Museum)

This abandoned or knocked-out Befehls-Jagdpanther was armed with a monobloc 8.8cm gun barrel. Three 8.8cm rounds are lying on the ground next to the left track. Zimmerit was not applied to the part of the sponson where the spare tracks were stored. (Patton Museum)

The same Befehls-Jagdpanther seen in the preceding two photos has a small pipe on each side of the left tailpipe, introduced in July 1944 to help cool the left engine-cylinder bank. A shield for the star-antenna base is on the upper left rear corner of the superstructure. (Patton Museum)

The 80mm armor of the glacis of the Jagdpanther was not proof against Allied antitank shells, as is clearly indicated in this photo of a knocked-out vehicle outside of Hargarten, Germany, on 15 March 1945. A tank destroyer of the 899th Tank Destroyer Battalion, 9th Infantry Division, First U.S. Army, placed an armor-piercing round on the glacis, leaving a gaping hole to the right of the Balkenkreuz insignia. There also is a large gouge on the right front of the mantlet and a hole through the right track from artillery hits. (NARA)